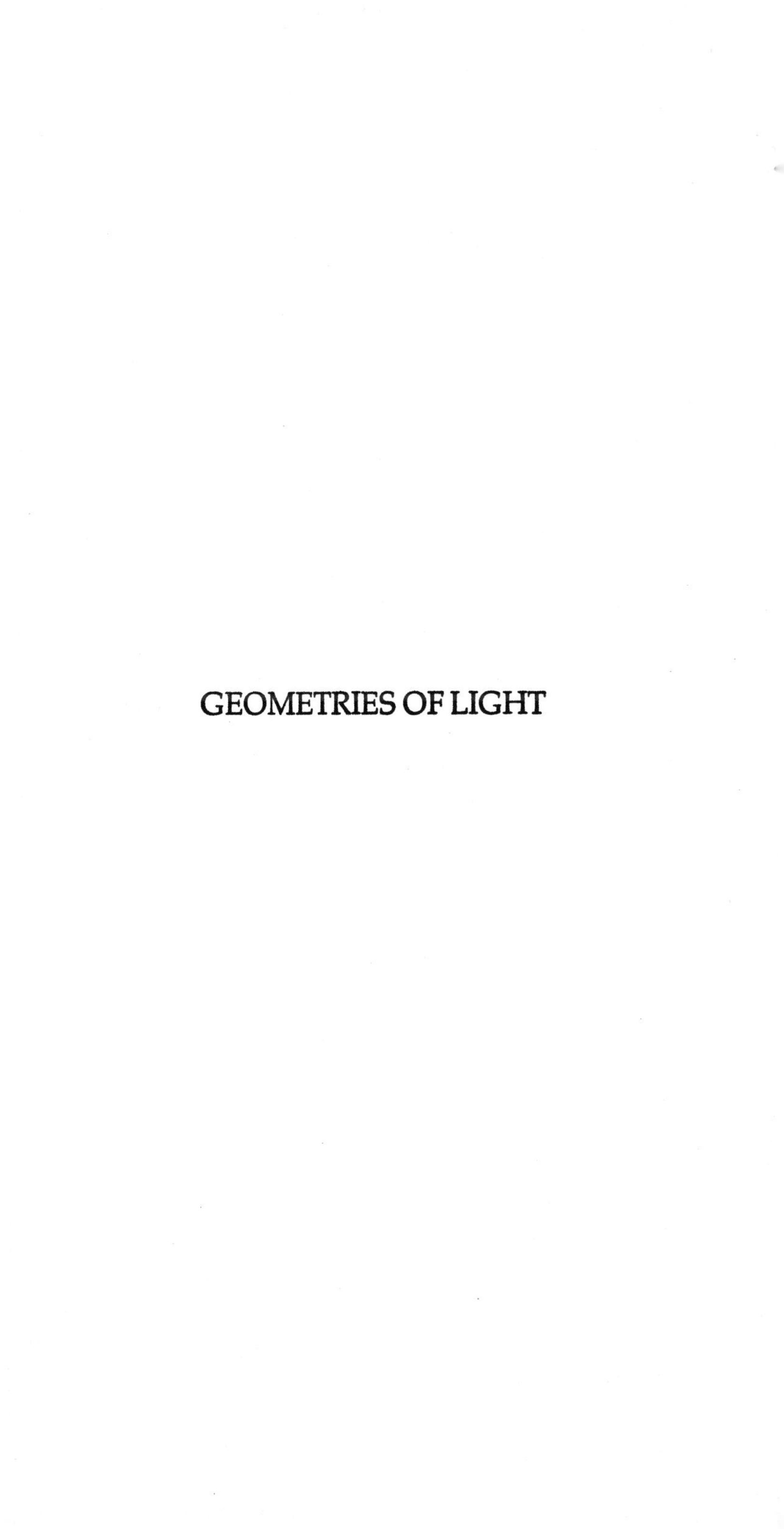

GEOMETRIES OF LIGHT

Other works by Eugene Warren:

Christographia, Ktaadn Poetry Press, 1973
Rumors of Light, Grafiktrakt/1, 1974
Christographia 1-32, Cauldron Press, 1977
Fishing at Easter, BkMk Press, 1980

Geometries of Light

Eugene Warren

Harold Shaw Publishers
Wheaton, Illinois

These poems
are for Rose,
wife lover muse

Acknowledgements

Poems in this collection have appeared in the following magazines and anthologies:

Adam Among the Television Trees
Aux Arcs
Being Blue Book No. 2
Christianity and Literature
Christianity Today
Eternity
For the Time Being
Good Seed
Grafiktrakts
Ktaadn
The Country of the Risen King
The Mennonite
North Country
Ozark Review
The Poetry Bag
The Reformed Journal
Sing a New Song to the Lord
Sou'wester
Webster Review

Cover Photo: Gary Irving

Printed in the United States of America

Library of Congress Cataloging in Publication Data

Warren, Eugene, 1941-
Geometries of light.

(Wheaton literary series)
I. Title. II. Series.
PS3573.A772G4 1981 811'.54 81-2067
ISBN 0-87788-300-9 (pbk.) AACR2

First Printing, May 1981

1
Towards Solstice

2
A Conversion

3
Bareback in Kansas

A Fore Word

Is poetry rational? *Does it* make sense?

In seemingly synonymous terms resides an ancient recognition of different ways of using our minds. Words like sensible, sensitive, sensory, sensual, *and* sense *come from a root meaning to perceive or feel.* Rational *and* reason *have at their base a concept of planning or reckoning (counting). Modern brain-wave detectors have located two distinct modes of mental activity in different halves of the human brain. So far the neuroscientists confirming the dichotomy have done little to help us become "whole-brain" people. Most poets aren't much help either.*

Poets and lovers of poetry are probably sensitive "right-brain" people who see a whole picture at once, value their overall impressions, and act on them. Sensing the rhythms of life, they tune in. Perhaps people who dislike poetry tend to be rational "left-brain" thinkers. For them, getting a true picture requires a tree-by-tree examination of the forest. They check out life's rhythms to see if a pattern is discernible.

A few people seem able to think both ways, through facile alternation if not actual integration. But in spite of C. P. Snow's early warning, the "two cultures" based on the two modes of thought, separate as ever, grow increasingly antagonistic. Poetic types sometimes put down "linear thinkers" as being machinelike creatures insensitive to human feelings. In turn the "fuzzy-minded" are regarded as irrelevant mystics estranged from the world's objective reality.

Is such a rip in the human fabric evidence that something has gone wrong? Recall the Biblical insight that willful human rebellion has wrenched apart a good creation: alienated from God, we are divided within ourselves, at war with each other, and a menace to our environment. Can poetry, or any other human endeavor, break through to renewed wholeness? There are both poets and scientists among those who see God's overall strategy as the one hope of gluing things back together, believing that his Son has already begun the reversal. Author of creation yet an unspoiled partner in humanity, Jesus Christ is a believable agent of recovery.

Even those who cling to that good news are affected by the same cleavage that spoils everything. In the interim before all is made right, a great interplay is necessary. We must be creative to play a redemptive role, yet our creative imagination must be ruled. To act caringly toward others is the shiningly simple rule. The half-brained on either side are called to reunite joy and responsibility, "on earth as it is in heaven." Scientists as well as poets can praise a Creator who playfully designed the kookaburra along with the dove, the hippopotamus and the horse. Thinking and feeling, work and play, are all to be orchestrated in harmony with the divine pattern.

Is it significant that so much analysis should flow from "Geometries of Light," this collection's title? The phrase (from "The dance of shining") has done its poetic work well, sticking in the mind to generate ideas and images. Under its influence, Eugene Warren's decidedly unmathematical poems can be seen as "geometries" in the original sense of that word. They take the measure of earth, our habitation. Earth? Yes, a lot of literal soil occurs in these poems, with rooting and growing and ever-changing mid-western weather. The weather shapes the landscape and warms or chills its inhabitants, who are, as Scripture says, "of the dust."

It is said that Egyptians invented geometry because the flooding Nile kept sweeping away their landmarks. A millennium later, Greek minds made geometry less utilitarian and more abstract. Warren's poems are metaphorical geometries in that refined sense, too. They examine properties and relationships without, like Plato, "reducing everything to geometry." They hold up to the light a point fixed in history or a turning point in a life, a line of reasoning or the various shapes of love.

Eugene Warren's poems are always careful constructions (a geometer's word) in which imagination knows its bounds but seems unbounded. Incongruities placed side by side become, as a mathematician might say, congruent: dawn is a "megaphone of silence." "Screw tin eyes" fits as perfectly as the "scrutinize" it replaces. All poetry is word play and this poet takes his play seriously.

To "play along" with a serious poet takes some work, though—just as those who would walk with God must adjust their strike to his. Grasping a poem's intent is more important than understanding everything in it, however. If, on occasion, we seem to be looking at splinters of thought rather than a finished carving, we can still absorb the poet's experience. Good poetry, multi-leveled in meaning, is sensible *even when one cannot* rationalize *every phrase.*

Eugene Warren is a fine poet with much to say. He has tasted the redemption that gives purpose to creativity, that puts the right things together. In an earlier collection, Gene wrote that he means his poems to be "Christographic" even when Christ is not their immediate focus. Sometimes his poems merely set the stage or hint obliquely. Sometimes they ask a question or cry out in joy or dismay. But we know that this poet can be trusted, because in key places it is so clear that he trusts God. At the very beginning he tells us what to expect: "when I rise/ it is in him/ word, voice/ Lord." And at the end, in "Bareback in Kansas," loyalty to the lamb of God pounds into rider and reader alike as a mare "lathers the wind" and a wonderful poem thunders across the page.

Walter R. Hearn
Berkeley, California

What does being a poet mean? It means having one's own personal life in quite different categories from those of one's poetic work, it means being related to the ideal in imagination only, so that one's own personal life is more or less a satire on poetry and on oneself.

Kierkegaard, *Journals*, tr. Walter Lowrie

1

TOWARDS SOLSTICE

My bones learn

what marrow
can my bones have
other than the word?
that thin core
of voice at the center
of all creature
whatever

when I fall
it is out, away
from that center
who speaks
and I am

(and I am,
O, driven on the wind
of that light,
as shadow blurring
into distance)

who speaks
and I am
and have chosen already
to flee that voice,
and have already
been chosen
to hear that voice
and return
O Center, to thy breath
that hovers
and wombs my chaos
into heaven and earth

what song
can my bones learn
other than that

groan resounds within,
and they are turned
to its solemn joy

when I rise
it is in, drawn
by the voice
to rejoice at the core
where I am spoken
when I rise
it is in him,
word, voice,
Lord

Silence singing

we move through
lays of heat,
ballads of humid air

mulberries rot scattered
among walnut sprouts,
squash blooms condense to gourds

six green pears are hard knots
of transmuted light
high on the tree—

the stillness of that light
a calm of song

The streets of the garden

The voice speaking
shapes silence.
The poem unwritten
moves arm & hand to reveal
its choice of line,
& the written lines shape
the blankness of the page.

The True man loiters
in a glass of water
cold from under earth,
clear as the grain
of time-licked stones. The shape
of the glass echoes
the City's horizon, the circular
infinite that bounds these fields.

The City is light
that is stone,
the City is grass
that is flesh, it is all sorrows
of small animals
& large man . bound by the horizon.

The dust is cold water,
the dust is man, man is a map
of the dust's way home,
and his body shapes the fields,
its lines drawn by the grass
and the cousins of the grass.

These fields are the water
drawn from under earth,
and their glass, the horizon

of hills.   The turned furrows
are a map of the seasons'
plowing, days overturned for seeding.

The seeds are words
sung in the City's streets.
The City's map is the community
of concentric lives:
the plan of the Garden
is the City's intricate
& costly delight.

Song into storm

heat broken
by a light shower

dark clouds
moving swiftly

the pressure of storm
heavy on skin
the light
as though carved
parts
to let the raindrops
drop
swift and separate
cool and heavy

songs in the dark
hymns
the notes falling
in clean lines
swift and separate
as the silence parts
to let them
through

Psalm for sprouts

Growing grief in plastic cartons,
splintery boxes set in slack rows
against smudged windows

The pale green cotyledons
pry back dark crumbs of soil,
their thin yellow stems leaning hard
against the fading weave of sunlight
and the late, chill rain
trickling down the wrong side of the pane

Someone lend strength
to raise the stuck sash
or splatter the bland glass,
open it even splintered
to the blazing wind,
to the green & startling roots
that grow in storm

Christographia 33

Adam asleep,
Missouri mud
under a crust
of yellowed snow.
Between his elbows
his bowels sprawl,
rife with mercy
and chert.
This is the winter
he fell through topsoil
and ended prone
on bedrock.
Roots cramp his legs
to their ancient crouch,
his gardener's kneel.
His eyes are bulbs
unsprouted but clear,
his teeth linger
in his jaws,
hoping to tear
a new loaf.
All his futile
husbandry
is fatal, mercenary;
he cuddles stumps,
not Eve,
cannot yet breathe
with the seed
that will crack
his dirt and make
of hurt a humus
for bliss.

The dance of shining

shapes shift
to the dance
of intent,
transform forms
through changes
of angle,
mobile geometries
of light
flowing up
from the hidden core

under the fields,
the mute light
takes stone
for flesh,
rises through root
tendril
stem
branch & flower
to be a liquid shining
in all things,
variously robed twin
of the naked sun

Song for the Santa Fe

this evening,
cool, dim,

we live
in caverns

of light.
Listen:

autonomous
sounds brown

against
storm-blue

stray sounds
overlaid

by a whistle
like an

Iron Cross

A white arc

the tomatoes
the vines
the red mud
the weeds & grass

the green tomatoes
made of light
& dirt
& water flashing
in a white arc
from the hose

the vines tied
with cord
with flowered strips
of old cloth
against white
pine stakes

the vines
the small yellow
flowers the young
tomato a tiny
germ of light

the red mud
soaked with water
the weeds clipped
& pulld   the grass
drying brown

the yellow
flowers the hard
green fruit

the seed-filld light

What the bluejay does

all the hosts of Sun

enter the room: mind bends
around the glass & fails
to comprehend
daylight on a child's elbow

•

motorcycles in the yard talk
to grease: steel puppets, they're

pulled through air, read asphalt comics

•

over the windowsill a vision floats:
sexy rain coating leaves & concrete,
rushing gravel down the gutter

what the bluejay does is private
in a cape of wind

Mutton songs

6 am alarm clock,
now noon, this still heat.
        Sky clear until
        eleven, then
        a few clouds,
        scattered,
the force thereof.

·

Canadian thistle,
light lavender fuzz
extrudes from green
diamond-patterned bulb.

·

That single
stunted elm
along the road,
a blob
of shadow, dark
dribble on
the one sky.

·

Talons of
the hawk
become vine, reverse
the process,

O

·

I bounce
through the worlds
in a blue truck:

Take heed!

Wet weather

In rain
I think of passing
galaxies

among the trees;
evening birds,
walking on the sun,

hiding day
above their stormy
feathers—

these cut stone
surfaces
left rough

or polished
by slow volumes
of air.

Eden's child
a prose poem

Before they disobeyed God and were driven from the Garden, Adam and Eve lay together, and so Eve was pregnant when the angel sent them into history.

Eve wondered about her unborn child: having been conceived before sin, would it be sinless or would it share its parents' burden? Or would the child unite in a mystery both its parents' guilty exile and their forsaken innocence?

The time came at last and Eve was delivered of the child with much pain and anxiety. Adam, the first midwife, had laid the tiny form at Eve's breast before they realized it was stillborn. The infant was of such beauty that they could neither look at it nor turn their gaze away.

Tenderly and sorrowfully, they examined the small corpse; on discovering the stumps of unfledged wings at its shoulders, they were shaken with a grief beyond even that of losing the Garden.

Laying the child down, they began to compose themselves to its funeral, but it melted away like snow in the sun. In a few moments, all that remained were the two ivory rings which had been the sockets of the child's unopened eyes.

Adam took the right ring and slipped it onto Eve's finger; she took the left one and slipped it onto his. Thus, for the remaining centuries of their lives, their hands were graced by the bones of their first child's lost sight.

The unseen appears as real shadows

the melted snow remains,
darkening the ground,
greying the sky

even what disappears
remains,
melting down to roots,
rising up to sun

the vanished snow returns
as new grass,
as february clouds
casting shadows on the hills

the images also
can be history,
seed swelling in grey mud,
new roots tangled
in the sun's fierce eye

February 24th—it's warm & blue

February 24th—it's warm & blue,
wind blundering every-which-way.
The last patches of ice and snow
are holes through time and space,
gates to the void beyond,
the seething vat
of which this world's the froth.
I sit, sore-throated,
and submit to the gusts of speech
that shape the contours
of this shaking world.
Two weeks ago, starlings
on bare branches
fluffed feathers against the cold,
swollen quarternotes of winter tune.
Now it's 50 degrees and sunshine,
rays of light pass
from their feathers into flight,
breadcrusts rise in vectors
drawn to the sun.
Beneath the carpets
of fraying cloud,
stumps of star-fire blaze
in the afternoon, joy so deep
it enters my feet
and gives them dance.

Roses ascend morning

The light becomes grass
and shines upon itself in form

The valley parts mountains
of evening and morning,
returns mist to the sun,
a rainbow bridging through rock—
each leaf a step toward the light
that waits beyond the crest

On porches of flesh
the ancients cross, step,
and turn, hands joined

Now, I step to you
on the pitches of night,
we touch, hold;
changes of key bring light
to our flesh

The Timbrel quick in the dark
takes our feet
and sets them in time

The Dancemaster burdens the tree
with darkness
which calls us back
to the turning

Roses ascend morning
and break the air open
to this song

Annual rings

cakes & sticks
of incense
fume & mist
in dawn's censer

blooms
of ripe smoke
rise
& hover over
long grass blades
sheathed with ice
(each wintered stem
ablaze
in morning sun)

a muskrat breaks
his winter sleep
to stir
the muddy waters

the smoke rises
through
the vaulted distance,
a grey & fading angel
in the sun's presence

Towards solstice

dark and cold
evening comes

fall, 3 weeks
from equinox

moves down
to solstice

the sun at
the south gates

snow on the high plains
ice in the fields

flowers and mandalas
of frost on windows

dark and cold
evening comes,

winter comes, the month
is half gone

and Rose, whose name
is a mandala

counts the thickness
of another year

a thirty-fourth ring
in the bole of her being

Dark & shining

still held stiff by winter
the vine-sheathed dogwood
flows toward bud

the empty snail-shell,
thin & grey
glows within

gravel & stones below
rumple the river's surface:
ripples & shadows trace
the current's force

my body flows
through self, channeled
by bone blood & childhood

the rumpled currents
of awareness reflect dark
& shining stones below

The angels shake

Buried angels strive to rise
against muscle-tension
clamping terror to bone. The stones
of their eyes fracture,
chips of limestone exploding
from the bluff.
Dogs in the shadows
gnaw feathered clots;
the angels crawl
through sealed arteries
to the heart.
The flames of the campfire
taper to darkness above the lake.

Hidden children cry
in the strata of the bluff;
their tears seep to its face,
vanish into green moss
on the angels' skulls.
Straining upward,
the angels shake the fire,
its spilled embers
scattering among the dogs
that gnaw and slink.
The lake tilts its wrinkled face
to mirror the one
who makes his angels
flames of fire.

Poem concerning time

for Jim & Mary Lee Quinn

3/14/69

There's never enough
time, for peace,
talk, relax
ation; to hold words up
to the strong, sweet light
of an evening together
to see what moves inside
the shadowy envelopes of sound.

When the elbow died
it was reborn
as an august grasshopper
leaping into the sun.
Or, a child crying
at night / is to be called
by no names   but
his own.

(As our own
names leap out
of places we've
named, renamed:
Wichita, Denver,
Kansas City,
Emporia, Mexico
City, Puerta Vallarta,
San Francisco,
Hobson Star Route . . .

names placed like stones
on the loose pages
of folly's folio, or
the quibbled quarto of events
half-lost)

part two.

A fence is made 1st
by placing the postholes,

cutting thru brown
matted grass, clay
mixt with gravel, gumbo
or rich, dark earth;
at last, 3 ft deep, the hole
begins to fill with water.

At that depth.

And the holes must be spaced
in as straight a line
as the eye permits
(sighting from corner to
corner of the field defined).

The post then fits
its socket of earth, bedded
among roots, pointing
straight at the center
of the earth.   What is

left is uncoiling
the wire, making it
sing between
the posts.   At that depth.

And when the fence
is finisht, it will
support a post if it be
broken from the ground
or burnt away, the charred
stub hanging on the wires.

Order.ly)

The images, at that depth,
damp with stone,
butterflies burrowing
where moles have no subway.

music/music/music

part three.

At the bottom of my lungs
the Bluejay hides,
slicing alveoli like
(French) birthday cake.

Frag.ment.s)
of a fence I built
15 yrs ago
remain.   We build fences
of talk, not to divide but
because a fence has
2 sides. If you strip
the bark from the posts
the wood is new & clean,
shiny & wet but quickly splits.

Season yr posts.

The corner of my father's
section was marked
with a limestone block
7 ft tall, 2½ ft sq.
& from the north face
to the east, someone
hand-drilled a hole
7 in. long, through which
ran a heavy, rusty wire.

North & east,
those 2 faces, the rust
dribbled from wireholes;
all 4 faces stained
with rain & birdlime, but

the corner *there*
for a 100 yrs
(only slightly tilted SW).

When the tongue died
it was reborn
as a cat hiding in lilac
bushes, waiting for
the family dog to screw up.

And he did,
getting 4 regular
clawmarks, a music staff,
on his cheek
(the 5th line being
his startled eye).

There is never enough
time, always too many
words. If yr eyes
match yr vision,
it's sufficient.

This entangled season

Stone petals unfold
in Spring twilight.
Tattered cloud-sheets restrain
the tumbling stars.
There are centuries between
the leaves of the crabgrass.
This entangled season
confuses: 1 walk across the street
to look back at my absence.
Birds fly through evening,
carrying in their beaks
coins struck
from the ashes of my heart.
The world in its silence
is too many for me,
despite the brief light shed
by image struck against fact.
The square of earth
lies within the circle of heaven,
man a pentangle inscribed inside—
a cyclist in the rain,
cubism of the soul.

Imago
Mark 8:22-26

I was burdened
by the wind,
sun split my eyes—
"Find me, find me,"
the hawk cried.

He spat,
a white,
glistening amoeba,
sprawled in dust,
thin bubbles
trembling in wind—

"Blind me, blind me,"
the spittle cried,
and gathered dust to it.

I saw men
with green fingers raised,
walking like trees,
shaking
with light.

A dawn

The Eden
of stone
shakes
with light.
The sky's
field is white
unto dawn.
The sun's
blade scythes
across
the hill
& cuts
the fog
from its roots
in water.
My eye
is level
with the
sun's
weight
bending
the horizon.
The Eden
of light
is firm
with stone.

Season

a year, a day
the ground-tone of season
the descant of weather

the prince of dreams holds court
in a fortress of snow illumined
by the moon's other side

when we are there
our eyes see below
light's habitual speed

that which holds us in place
is season, the deep shiftings
of earth in its turn

that which keeps us moving
is place, the opening field
where all contraries dance

when the sun rises
he draws us up
droplets from a shadowed pool

and as we transit the sky
we are a prism to refract
his seven-fold glory

Ode: entropy & Easter

that all things wear out, break down,
erode, crack, shatter,
sag, splinter, break;

due to neglect
the rooftree gives way—
and there is no way
to avoid neglect
something is always left undone
something always overlooked

if we knew when
the thief would come,
cancer grow,
the bombs fall,
we could take reasonable precautions
but we work against night and decay
with little light,
and in ignorance of the next moment

the seed must die to grow
and yet our life—
scripture says it—
is as brief and tenuous
as the wild flower trembling
in every breeze, scorched
by the sun, clipped
by the frost

all things run down to dust at last,
they crumble and scatter, are lost
the wind chimes clatter
against Rose's sung alleluia,
the guitar's throbbing chords,
as I reach with ink
for an affirmation, seeking

a light seeded and rooted
beyond—beneath—above
the light that is only sun

how easy to say
"he is risen, it is Easter at last
and darkness has lost"—
yes but harder to say lightly
against the weight
of a body riddled with cancer,
of a child tortured and murdered
of the twist of the spine
that makes walking a cacophony

Easter can be true only
if the Cross was truly
the death of God
and Man in one body—
sung alleluias flower
only from that dark root
of final disaster
when all seeming hope is lost—
then can the dead rise—
and praise be alive at last

2
A CONVERSION

The splice

the news.
the wind in the mimosa
rattling the seed husks.
the glass chimes
melted rings of green and brown
clang in the wind.
the apparatus of the flagpole
clanks.
dim lights across 2nd street.
across first.
space heater's metal creaks and snaps
as the burner burns.
the marquee of the Ritz.
rows of streetlights.
a nerve, one end in the world
the other in the soul.
the news.
papers piled for recycling.
ecology of print.
food-chain words.
phrases. sentences.
language feeding
on itself.
an endless belt.
but you can see where
the splice of silence
holds the chain of words
together, where
voice is new and abrupt,
one end in a throat, the other in an ear.
one ear in a dictionary,
the other in a furnace.
the news.
the body's sounds, folded
upon itself, suspended
in a vat of light.

a wad of frozen beet greens
thawing on a board
a dark vegetable riddle
speaking too slow
to be heard, roots in a yard,
leaves in a kitchen,
my body the splice.
the news.
we grow in and to speech.
the wind chimes.
glass against glass.
metal against metal.
one end outside,
the other in.
the body is not a box,
the person not a space enclosed.
in perception's thermocouple,
we are current.
a hail of leaves on the wind.
skin is the organ
of cosmos and self.
the news.

Antenna flags

antenna flags
go grey slowly,
fading to colorless
rags, soft

as clouds, smooth
as nipples, silent
as civilians
in a mass grave

The assassin of God

The assassin of god
was a heart sewn
on an empty sleeve,
a foot marching double
file on a one-way street.

The assassin of god
was a man-shaped
blankness, sucking
at the world's edges,
decaying sunsets.

The assassin of god
went to church, rang
clapperless bells,
remembered his mind
in time to leave it.

The assassin of god
was a pillar of fire
in steel days, made
his life a movie,
and smasht the projector.

The assassin of god
cut his wooden tongue,
nailed himself with ice,
walkt into the harbor
of eternal absence.

The assassin of god
ambusht a metaphor,
drew a target of pale blood,
took a torrent of data
and ate darkness raw.

The assassin of god
murdered a shadow
and called it natural,
ran to the window
and leaped into himself.

Hunted by dark mouths

Freud & Marx & Darwin

the angel on his Harley
hunted by dark mouths that ride
the wind   what's sown

knows no reason   pears rot
Tooth shatterd in the moon's skull
Moonless nights unroot the Changes

iceglazed blooms
fire in the lake, rush
& turmoil of steam, the deep
burning   suntiger drownd

Light submerged drowns roots

(Keep
the whirlwind from
these people   they

are helpless

State pen

Electric gates spring open
a lion's mouth

Beyond
stone & mud,
drooping sunflowers
grey & black seeds
sprinkle the yard

Immense zinnias,
marigolds;
their blooms blacken
like old blood

Their stalks
& heavy leaves a script
scratched
into boredom's wall

The mimosa trees
let their leaves fall,
syllables of a sentence
passed on the season,
winter's grey
& solitary
cell

The bantam spoiler

Orange letterd on black:
WALLACE
enlivening the rusty
chrome.

Dusty backseat
children ride
grimly, their blond
hair tangled
in Federal laws

—the travail of WASPs whose
rhythms are
not their own—

dulld by turning back
to the dumb drums
of death's unrhythm.

As if to speed
their passage,

in the rear window
a Bible
flutters blackly,
wings of
                    outrage

in the dust
of daily travel.

A survivor of Babel

He flexes his lips,
tenses his jaw,
clenches his tongue & uvula
but makes only silence—
silence & the rasp
of tissue, pop of spittle

Somewhere his voice
is filling a stranger's throat
& mouth,
words he should speak
are clearing a stranger's lips—
& only its distant whisper
tickles his ear

An evening

The spider's tomb
is another edge
of the same light
that baffles my eyes
in their urge to see
through.

Fragments of image,
partial syllables,
float & whirl,
motes in the throat's storm,
hoping to fall into
some pattern that
sings.

The ink threading
through this grey pen
sews words on the page,
hemming
the ragged edges of my
silence.

The spider's tomb
is a granite ear
in the evening sky.
My children cry,
"The moon is out,
the moon is
out."

Eleven splinters

1

the deliberate death
of whoever will not face
the intricate geology
of our flesh

2

"take it

slow"

must have time
to experience
time

art is short
& life is long
ing

3

the silk flow
of mother's milk
catches in folds
of worn cloth

4

an articulate despair
starting from obscure intuition
of the bone's sentiment,

secret visions seen
by palm's wrinkled eye

5

her nipples sucked
by a thousand eyes

6

unsold hayrakes stand
in waist-high clover,
new paint
stiff with rage

7

this key
found in mud
is no use
without its lock
except to open
imaginary
gates

8

an unworded grace:
these blue
morning valleys
cold stars riding
in the boughs
of dying oaks,
letting our hopes settle

warm dregs
to the sun's rim

9

some Schoenberg
discords amid am static
car radio pops
when lightning flicks
the clouds

tangles of violin
& bass viol
torn metal signs
swinging
in a manic wind

10

How's a young man
to keep his way
clean?

this world lodged
in my eye
like a word which
in speaking
I forget

screw tin eyes

11

a special grace
in this turning
piece of air

this tapestry
moon image
of a further
place

Fishers at dawn

It is the peace we refuse
rules us   .   roild ears of wind
globular discontents searing
the edges of our journey

When walking by water, to hear
plash of fish leaping
waves licking boulders, the creak
of piers empty with dawn

The prow turnd windward, the boat
        slips over the waves,
                each by each,
        as the wind rises,
the whole surface of the lake
moving with wind
                                & against
         that movement, the oars
& the oarsman's muscles,

a field of tensions in water & flesh

When the line is cast & the bait taken
the depths are sounded
                                        against the strength
                    of the line
         & the angler's skill   .
                    a peace refus'd

The way

Walls
divide / doors
open

Roads
lead to
doors in

walls
that divide /
Windows

pierce
walls with
light, gold

& green, red,
silver, blue &
white:

scenes
of roads that
lead

between
walls crossed
by

shadows
of travelers,
one,

two,
three, a
dozen—

all
going from
door

to door / dividing
walls, pierced
with light.

Letter on itinerancy (1742/3)

Sir,
these two angels of the End,
James Davenport & Benjamin Pomeroy,
came pretending visions,
as two witnesses to condemn
the "blind & dead" ministers hereabouts;
"a multitude flock'd after,"
singing in streets & lanes
and the like unruly actions.
We heard them shout
of the hell-flames licking our cheeks
& shudderd.
Also, they made holocaust
of books of heresy (such sermons
as the unconverted preach);
& stripping in a most public place,
then made a promiscuous heap
of "Cloaks, Petty Coats and Breeches,"
to burn those idols of thread—
"thereby stumbling the Minds of many."
Their bodies being arrested
yet they made a mock at our laws
and government, speaking
of the sudden dissolution in flame
and, as false Elijahs, calling
fire down to consume the sheriff.
No fire has yet fallen
thru their empty words
& Mr. Davenport lately abjures
his former works of madness.
Now those voices of shout & singing
from hedges and fields
beyond our meeting-house

are silent; & we await the fall
of fire on our quiet altar
to illumine our reasonable religion.

Yrs &c.,
Anti-enthusiasticus

A conversion (1741)

Nathan Cole quit his plow
& hurry'd to hear Whitefield preach
& found his religion useless ash
before the stern words
"election" "grace"

"Hellfire   hellfire
ran Swift in my mind

"And while these thoughts
were in my mind
God appeared unto me
and made me Skringe: . . .
and I was Shrinked away
into nothing"

And in that nothing
Nathan Cole burst
with light, found
Farmington, Connecticut
ablaze with matter for praise:

all walls & common fences;
weeds, trees,
                                        and vines in special;
stones he'd earlier envy'd
for lack of soul—
now were new tongues to hymn
Election's glory,
Jehovah's sure salvation.
Selah.

Darkness

The darkness exists;
it does not have to be imagined.
If I forget it,
it remains, a stain of shadow
under my feet,
at the nape of my neck,
leaking through my heart.

When we would lock the darkness
away from us with facsimiles of light,
we only feed its falseness.
Striking a match
on the wall of my flesh, I see,
after the pop and flare have dwindled,
after-images of my face
receding into night.

The darkness exists
and is more than our ignorance of light
and is more than the shadows cast
by our pride and fear.

Yet the true star is kindled,
a straight blaze of sun
before which darkness flees
and gathers itself
into its own shadow.

Sight in bloom

fire spilt from the altar
burns eyes in the fabric
of my ease

crumbs of flame,
drifting earthward
on the wind
that measured chaos,
burn straight through
my layered costumes
and seed my heart
with blooms of light

"As an army with banners"

The hunt:
handletterd
vellum parcht
in Imperial flames,

the written word gave
a flickering,
smoking light.
papyrus
scrolls leave
ashes of revelation to darken
the Emperor's mind.
Burnt words spread
the Word.

Anointed
with Nero's oil,
the saints
were candles of flesh

but their burning eyes
saw the Beast
cast down into
endless fire & dark.

"These atheists
are foes
of man, god
& state.
Both them & their fool
of a christ."

Ignatius, Polycarp,
Felicitas, Perpetua,
Blandina, Stephanos,
Petros & Paulos:

blood seeds dropt
in the Roman furrow.

"Into which angels would look"

blessings to shrive
the angels' eyes
who cloud about us
wings riffling
                    among the leaves
(sigh of the wind
in a forest of living bones)
making do with thin
dry loaves moist
leaves that float
on the air
wingtips shave & carve
the light roaring down,
make forms that blaze
angelic
            among the clouds
that gather to gaze
at our amazing bones
all split
            by urgent buds

All these breads

all these breads—
matzo, rye,
tortillas, soft indian disks,
unbleached wheat—
broken, torn, snapped, crumbs
floating down from soft loaves
or popping up from the sheets
of perforated matzo—
these many grains
grown in red soils, black loam,
grey or yellow clay,
roots of wheat and oats
and barley and rye
probing dirt & rain,
the slender, parallel-veind leaves
arching in sun or lying
straightend in a strong wind—
crusht, ground, rolld, sifted
at last becoming
all these breads—
one diverse loaf passing
from hand to hand,
dying into each mouth,
sprouting a new
& shining grain

3

BAREBACK IN KANSAS

Matthew 5:30

Because it offended me,
I lopped off my right hand
and dropped it behind me
into the shadowy noplace
where the Adversary is said to lurk

It fell as a root
and burrowed thumb-first
into the blind field,
sprouting fine white tendrils

Its chill blossom, a crown of fingers,
wavers in my sleep,
the petals cold and blue

I pluck that bloom for candles,
lighting them with a knife
dipped in blood and water

The light they shed is a web of shadows,
on which that severed hand lurches,
a maimed spider,
dribbling behind it a thread of regret

Better to lose that crabbed part
than to find at the end
my whole body grown to a stalk of weed
to be plucked up and burnt,
a candle of desire burning itself to naught

Shoving free

impatient
uncharitable
I spin my wheels
in the slush
of February 1st

thirty-two degrees
of Fahrenheit
we can't back into
this parking place
and can't
drive out of
as much of it
as we've gotten into

the right corner
of our front bumper
rides under
the left rear bumper
of a bronze Continental
which, if it weren't there,
we could drive away from

a kid sets down
his backpack
hand-gestures me
as though we speak
different tongues

he shoves
you shove
I shove
another guy comes over
and shoves

our Buick at last
backs free

and

impatient
uncharitable
I drive you
home

In the rain

my rage in the rain
deliberately stomping
through puddles
to wet my childish way

muttering bitterly
in my beard, letting
frustration with circumstances
sour my love
and violate my heart

and when the current of grace
rose in my heart,
the appeal to mercy,
the shame
that I at first resisted
and shut my mouth
on the words
that call out to the one source
of ease and forgiveness,
ah, mercy,
mercy

my rage in the rain
mirage in the rain
marriage in the rain

The dumbfounding

for Margaret Avison

The song-seared tongue
can be dumb
with melody   as bone
stretcht beyond breakage
drums the cadence.

The soul's Herder
sings us several into one,
His, as we gallop,
canter, trot, ramble,

a flesht and many
temple,
into the fold where
our tongues hold

song
as only leaves,
dumb with greenness,
hold spring.

Song

When the angel appears
on the shores of sleep,
we hear murmurs within
his foliage of eyes—
at that shore
where sleep eases up,
crests, tumbles
& slips back to deeps,
leaving on the slick beach
stones worn as clean
as the boneprint
of the angel's feet—
at that shore
the darkness rolls up & drops
& in the suck of backwash
the angel appears,
in his sixfold wings
a surf of light, in his
silent hands
the vowel of dawn.

The sleeper walks easily

The sleeper walks easily
down the hidden road
to a dark river in the heart.
On its further shore,
denied selves cry & flicker;
the dreamer brings no white bulls:
spills instead of blood ichor
drawn from the narrow vein of waking.
Crowding close, the ghost selves
gulp the dreamer's offering,
then caper with fading shape.
Waking, the dreamer moves
in a shadow-cloak ghosts wove
of his fear & hope, a flickering cape
of loss lined with desire.

Waking near midnight

A grasshopper
carved of snow
leaps
into the harp
of held desires

Its thorax twitches
with the certainty
of noon,
scintillate tweezers
of light
that pluck shape
from its freezing matrix

Dreams we cannot recover
before they melt,
living spokes
of the mind's wheel

The sequence of dreams

The sequence of dreams
has a sexual aroma,
mother's milk,
fall rain
stippling dust

Your breasts
wrapped in flannel,
swelling
with new milk,
their fullness
overflows,
soaks nightgowns,
bras, blouses,
traces pale wet lines
on my palm

I touch you
for love
where love's fullness wells
from the depths
of your body
& drenches
our night-locked flesh

Late March love poem

one red tulip
multitudes of jonquils
a tilted wisteria
tiny wild blue violets
I am delirious with her kisses
come, let us walk
along the tracks and sing small buds
the iris will bloom later
peonie's red stalks press upward
I long for coleus
the sun is a rose too hot to sniff
I am pricked by her touches,
yea, she causes my heart to pant
Oh Solomon, your proverbs,
your Song, your weariness,
your odes, I am comforted
with the least apple-seed
of her crisp, sweet flesh,
the softest syllable of her psalm
all our branches have buds
my desire is for my beloved
and her desire is mine
what sweet billows crown the horizon
yea, I will feed among her blossoms,
the tender grasses of her love
Ah, my sister, my garden,
my bride, Spring is here,
multitudes of jonquils
one red tulip

Body English

I spell yr name
with an alphabet of skin,
syllabic fingers tipt
on my flesh
                    all the vowels
we kisst onto each other
night & day   & consonantal
ticklings, pinches, tonguings,
caresses—

the syntax of our body
merges verb and object,
& no conjunction
is subordinate.

"My eros is crucified"
Ignatius of Antioch

we are each the form
of the other we love
skin no longer a seald shape
but an opening sheath

in our mouths one voice
in our lungs one breath
in our veins one blood

any least image
may a greater bear: we, too,
rise again
after dying in love

the inner curve of yr thigh,
cupping up
to life's tender gate
is the shore
of another world,
a coast brimming
with unsought glory

Before dawn

Before dawn,
birds make masses
of twitter and chirp
as we rouse in the dusk.
On my desk,
tangles of lemon thyme
scrawl toward the sun,
flecks of sap glisten
on the hyssop bush.
Outside,
the leaves
of the cucumber vine
wither, bees hum
in the limp, yellow blooms.
Heat stands listening
as waters gush from the hose,
soaking down and under
to the roots.

We touch a touch of rooms
and houses we've lived in
the vines and leaves
of our space.
Looking
for a door
in the solid blue air,
I find this clarity:
whatever pebble or flower
I bring you is no thing
but a knot of light.
We enter this.

Christographia 35

A spring bough to Rose.

Not by chance
the cock sparrow
treads his dance,
hen crouched
under him
on the narrow
icy limb.
Love is given
out of wintery
heaven: we,
in our long-wed
limbs, dance,
a slow & secret
step, tread
a measure narrow
but deep, find
our branch steep
with glancing light.

Christographia 38:
After Equinox

The clock's hands fall
& rise to midnight.
The wind relentlessly
attempts the windows.
Four seasons,
four elements,
the humours of flesh.
We lie together,
our body a glyph
for a deeper dance.
Late March rains
dribble from the eaves,
hiss under the wheels
of passing cars.
Our children sleep,
bedded in our marriage.
The outer doors shift
in their jambs
as the wind rises.
Midnight is heavy
in the rain,
the air dark with Spring,
foreshadowings
of seeds that die
to sprout again.
Four points of the compass,
four arms of the cross,
the bleeding center,
a rose of pain.
The needle dips & swings
as the compass tilts,
homes on North,
as the clock's hands
consent to midnight
and day shifts in its jambs.

We lie asleep,
belly to back,
arm over, arm under;
our flesh, having dipped & swung
has now settled true.
Four arms, four legs,
our body has,
a tangle of belief
we're learning to sustain.
Late March rains
dampen midnight,
flow into our children's sleep,
as they shift in their beds,
their weight balanced
against the world's tug.
I see no stars
as I shape these lines,
but hear, beyond the traffic
& the rain's thrum
& drip, the silence in which they hang,
pointing North, measuring midnight,
that silence the tonic
of a song we have yet to sing.

Thus we speak

thus we speak
into the megaphone of dawn
hear echo of sight
falling thru trees beneath our shadow
the wall is the law
the shout of shadow triumphant
as it divides figure from ground
body from motion
time from space
this is the night when seas overflow
with echoes of trees that shout
into the dark muzzle
of dawn, that megaphone
of silence
the wall relaxes
its stones slip free & coast
away to stand on end
among the trees of surf
rooted in the hollow cone
filld with echoes
of our sleep

Seed

fire
salt
stuck in the same
deadly rut
of usual words,
the personal cliches
that bubble up
when I "lean
and loaf and
invite
my soul"
and my soul
is surprised muttering
the same
old attitudes
over and over
fire
salt
the shiny
green
of plastic foliage
in a lobby,
any lobby
stuck
in sameness
in responses
as unvarying
as the internal
stimuli
and their external
correlates
the sensory bits
that mosaic
the appearance
of appearances
fire

salt
cold blood
red vinegar
in a corked bottle
a tray
of stars forgotten
at the back
of my large flat
desk drawer
jammed between
a postcard of
a Modigliani nude
and a plastic carton
of thumbtacks
shiny and
flatheaded
fire and salt
the tiny green sparks
of mineral heat
flare like
the syllables that march
through my throat
and vanish between
my teeth
fire and salt and
roses like blood
congealed
on a seed
of sound
salt-spark
fire-flare
cold cold cold
the tack's point
as a grain
of burning salt
stuck on
my lip

Psalm: The dark night

The Dark Night
is its own reward—
grit without loam
chaff without grain
shadows without light
shape without form
voice without speech
faith without grace

a face with no mouth
sings a song with no tune
an arm with no hand
holds a book without pages
a heart with no chambers
pumps blood without being—
the Dark Night
is its own reward.

Daylight

the talents of daylight
are to sing stone to dust

to carve leaf & stem

to wait still
on the rippling stream

to hover on the current
that pours through the gates
of the eye

there are no angels
brighter than dawn

Sleep song
for Howard Schwartz

The water flower's white leaves
carry a script
dark & cursive,
scribbled on fluid space.
Unconcerned for breath,
I follow those words
into the pond, their letters
waver just beyond my hand.
Folded in the pond's flesh,
I wake in a further dream
that flowers
on bright roots which burn
and are gone.

Within the concentric pond—
within the mottled dream—
within the bone cosmos—
heads of drowned grain sink
in the green depth
of my body,
that various pond
in which my days,
my doings,
have taken root
and speak stars underwater,
writing infant suns
across the chart of my nights.

Deathbed

The land behind the old man's eye
  falls clear and precise

into shadows etched
  on his flesh—bright rivers

bend from his bone, flare
  in the light before they curve

away beneath the bank
  encumberd with green;

beyond, on the sandy meadows,
  goats and antelopes crop the coarse grass

& feed beside sleek horses
  ornery with oats.

At the border of this land
  he lies, the old man,

peering out at the chromium
  utopia, the stainless light

that pours on his bed;
  his aged flesh settles

into the streets, forgetting
  rules of order—soft wet stains

spread around him, slow drifts
  of death rise to the surface

of his body and flake off white
  onto the cloth over him.

But within the knots and snarls
of his skin, stringy meat & bone,

lies a land—not final—but pure:
a group photo of all his days

enlarged to a newness
that absorbs him;

& he draws away from the white
uniforms & the pasty meals,

his feet tread again that land
for the first time.

"An enemy has done this"

The cruelty of objects
is their lack of urge,
they are heavy with lack
of innerness.
The plants thrusting aside dirt
to get to sun
flaunt their growth—
the thick, heavy heads of wheat,
the green flocks of soybeans.
The plants are not objects,
but are simulacra of flesh.

It is flesh
in the black cadillac hearse,
a cosmetic void of flesh
closeted in satin and steel.
The flesh is not an object,
it does not lack innerness
but is plowed & sown
with innerness.
Yet this flesh being hauled
is collapsed about its innerness;
it is darkened about the lips
and wears it hair differently.

The cars that follow the hearse
are objects,
as are the marble headstones
incised with letters and numerals—
these have no urge but rest unmoving
in the law of objects.
The plastic grave-flowers
are objects and not, as real flowers,
simulacra of flesh.

This flesh being hauled
in a black cadillac hearse
is now a simulacrum of an object
in its silence & stillness;
but the flesh still has urge:
it would fly up,
it will fly up,
from its dust,
thrusting the earth aside
to enter the light
and be remade without shadow.

Another birth

The mons shaven for birth
gleams gray;
her legs cocked in steel stirrups,
the woman heaves and sweats,
moaning into the mask of ether.
The baby drops, glistening,
into the doctor's tight, slick gloves.
Belly slack, womb empty,
the mother relaxes her jaw and sleeps.

The old woman's patchy gray hair
is worried into stiff bunches.
The bright flowers on the dresser
are repeated in mirrors.
Steel, glass, rubber.
Her lips fold against her gums,
morphine lowers her eye-lids.

In the clean womb
of Intensive Care, she writhes
with pangs of another birth.
Shaven at last of all that dies,
she presses open the gate of flesh.
O Mother: what convulsions of life
our deaths bring.

A signature

—went into the world like a shining knife—

they don't like the blood,
the rusty nails crookt
with gore
                    the hot gush
out of split skin

& the hot wind like boild sand,
a bath of grainy steam

the wildness, a black sun
& the shaking
is terrific
                    they don't like
the stench of death,
not even rich heaped manure
rotting for the garden

(delicate glaze of blossom formed
of dung translated into cells of sweet)

they don't like the body coming
back like that, wounds open
to sun & air, walking around,
they don't like not being able to vanish

Bareback in Kansas

The mare lathers the wind,
her mane streams like light,
my face is full of it;

I ride her like a lord of pastures,
a meadow in each eye,
stockpond deep in the center:

water down to mud, mud down
to limestone colorless
at those depths,
greasewhite until sun yellows it.

I am thinking of You
as her hooves bite the grass, spreading it;
I am thinking of Your face
bearded & serene, of Your eyes like the pond on a clear day,
a double depth cloudless;
I am thinking of the mouth in Your side
that spoke the fountain,
of the dark bloodcaked eyes in Your hands and feet weeping,
I am thinking that You loved me as I mounted the ladder
& shoved the thorns around Your skull,
I am thinking that the palms of Your outspread hands
watched me as I turned from the hill
& went laughing back to the city
to spill wine like blood down my throat
& tell whores of the Fool.

I am thinking of the spearthrust
that brought the fountain from the rock;
I am thinking Your dead eyes held my image,
I am thinking You broke the darkness
& came after me,
I am thinking You tore the weeds from my flesh
& sowed good seed,

I am thinking of the nails driven into Love,
I am thinking of the governments raising steel helmets
against You, of the nails of denial in our mouths,
I am thinking of Your look that changes,
of the Light that sweeps from Your wounds.

And the mare races through the pasture,
her mane flies in my face,
I lie close to her neck,
the speed of her gallop is not more
than the speed of Your mercy:

And I know that You loved me
though the hammer was in my hand,
though the spear was registered in my name,
though I laughed & taunted—

You did not crush me, You hunted me,
& the swift arrow of Your mercy
shattered the swollen ball of my selfish eye
& You wiped the pus from my cheek
& Your kiss blossomed my sight anew;

And I know that You are the lamb,
that You are the tiger;
I know that Your love stands against all night,
that darkness' king has known the temper
of Your blade & fled;
I know that none evades You,
that death is shattered on Your rock.

And I know that this mare will rise with me,
that You will touch body as well as spirit,
that the blossom will have its stem,
that Your city stands forever,
that the tree bears in season and out,

I know that You know my name
& call it,
& my answering is to Life.

If you liked this book, you'll enjoy others in the same series:

The Achievement of C. S. Lewis, by Thomas Howard. "Written with Lewis's own passionate power with words." —*Peter Kreeft.* Paper, 196 pages

Adam, by David Bolt. An imaginative retelling of the Genesis 1-3 narrative. "I think it splendid."—*C. S. Lewis.* Cloth, 143 pages

Creation in Christ: Unspoken Sermons, by George MacDonald, edited by Rolland Hein. Devotional essays revealing a deeply moving understanding of holiness and man's relationship to God. Paper, 342 pages

A Guide Through Narnia, by Martha C. Sammons. A detailed study of Lewis and his Chronicles of Narnia, with map, chronology and index of names and places. Paper, 165 pages

Images of Salvation in the Fiction of C. S. Lewis, by Clyde S. Kilby. Explores the Christian meaning in Lewis's juvenile and adult fiction. Cloth, 140 pages

Life Essential: The Hope of the Gospel, by George MacDonald, edited by Rolland Hein. "A book for those who hunger after righteousness."—*Corbin S. Carnell.* Paper, 102 pages

Listen to the Green, poems by Luci Shaw. Poems that see through nature and human nature to God. Illustrated with photographs. Paper, 93 pages

The Miracles of Our Lord, by George MacDonald, edited by Rolland Hein. "A better set of meditations on the miracles of Christ would be hard to find."—*Walter Elwell.* Paper, 170 pages

The Secret Trees, poems by Luci Shaw. "These are the real thing, true poems... they work by magic."—*Calvin Linton.* Cloth, 79 pages

Tolkien and the Silmarillion, by Clyde S. Kilby. A fascinating view of Tolkien as a scholar, writer, creator and Christian, based on Kilby's close association during the collation of the Silmarillion. Cloth, 89 pages

Walking on Water: Reflections on Faith and Art, by Madeleine L'Engle. Shows us the impact of the Word on words and ourselves as co-creators with God. Cloth, 198 pages

The Weather of the Heart, poems by Madeleine L'Engle. "Read her poetry and be chastened and filled with joy." —*Thomas Howard.* Cloth, 96 pages

The World of George MacDonald: Selections from his Works of Fiction, edited by Rolland Hein. "A treasure of a book—one to be read and reread."—*Frank E. Gaebelein.* Paper, 199 pages